U0195748

滬港名媛旗袍寶鑒

宋路霞 徐景燦 周鐵芝 編著　王逸 俞晨瑋 譯
宋路平 攝影

A RARE APPRECIATION OF THE QIPAOS WORN BY THE GRANDE DAMES *of* SHANGHAI AND HONG KONG

上海科學技術文獻出版社
Shanghai Scientific and Technological Literature Press

《滬港名媛旗袍寶鑒》

A Rare Appreciation of the Qipaos Worn by the Grande Dames of Shanghai and Hong Kong

編 著

宋路霞　徐景燦　周鐵芝

AUTHORS

Song Lu-xia, Jeanette Zee, Zhou Tie-zhi

翻 譯

王　逸　俞晨瑋

ENGLISH TRANSLATION

Frank Wang, Yu Chen-wei

攝 影

宋路平

PHOTOGRAPHY

Song Lu-ping

編輯委員會

Editorial Committee

目錄 Contents

（以姓氏拼音字母为序）

前言

本書是繼《上海名媛旗袍寶鑒》《中國望族旗袍寶鑒》和《中國望族旗袍寶鑒》（續編）之後，推出的第四本名媛旗袍書。

此書收錄了上海老旗袍珍品館近幾年來收集和借用的，26 位滬港名媛的 59 件旗袍，其中包括前香港特首林鄭月娥女士的 2 件旗袍、榮宗敬先生夫人榮劉惠秀女士的 6 件旗袍、宋子文夫人張樂怡女士的 5 件旗袍、齊如山先生的外孫女賀寶善女士的 2 件旗袍、香港著名教育家邊陳之娟博士的 3 件旗袍、寧波天一閣藏書樓後代范元芳女士的 7 件旗袍、香港著名兒童教育家鄧戴月華女士的 7 件旗袍、著名影星李麗華女士的 2 件旗袍……這些旗袍，都是名媛們當年穿過的，承載了她們特有的中西合璧的服飾美學理念，同時也是當時紡織、印染、設計、裁剪、縫製、繡工等技術發展的印證。通過對這些旗袍的介紹，我們力圖展現滬港兩地百年來海派旗袍的精華，與以前出版的三本旗袍書一樣，將為海派旗袍的發展史，留下一些真實的記錄。

上海老旗袍珍品館成立於 16 年前。16 年來，我們在眾多朋友和親友的支持與幫助下，走出國門，走向大洋彼岸，到世界各地去收集、訪問、搶救，老一代上海名媛穿過的老旗袍，目的是為中國的海派旗袍文化，留下一筆不可多得、不可再生、稍縱即逝的真實資料；為後人留下一宗優秀的旗袍文化的資源寶藏。

到目前為止，我們已經從紐約、洛杉磯、舊金山、東京、新加坡、香港及上海等地，從老一代上海名媛的衣櫥裡，收集到了 600 餘件老旗袍，並且已經形成了家族系列，如：宋慶齡家族、盛宣懷家族、榮氏家族、牛尚周家族、顧維鈞家族、徐士浩家族等等，還有很多滬港名媛穿過的精美旗袍。其中榮氏家族的 117 件老旗袍，已經無償捐贈給上海大學博物館，以期依靠大學博物館在展示、保存和研究方面的優勢，更好地發揮這批珍貴旗袍的傳承作用。

16 年來，在海內外向我們提供老旗袍，以及幫助我們獲贈老旗袍的老一代大家閨秀，已經先後有 39 人離開了我們。我們非常懷念她們。值得慶幸的是，她們雖然已經遠去，但她們的旗袍已經回到了上海。這使我們更加清楚地認識到，這項“老旗袍搶救工程”的重要性和緊迫性，如果不盡早地把它們從世界各地一一收集起來，海派旗袍巔峰時期的陣容，就將因為缺少實物而淪為紙上談兵，中國旗袍文化的歷史，就將缺失一頁。為此，我們感到十分幸運，我們在這些老一代名媛的幫助下，有幸親自參與這項非常有意義的搶救工程，16 年來，留下了此生難忘的記憶。

本書著錄的旗袍是我們“老旗袍搶救工程”的一小部分。它們的款式不同，面料不同，來源不同，神采各異，如今能夠在上海匯為一編，實在是一件非常難得、值得慶幸的事情。

展望經典、致敬經典、研習經典，最終目的還在於傳承經典。我們在本書的附錄中，展示了上海芝蓮福文化發展有限公司複製的一部分名媛旗袍，希望得到專家們的指教。

這些成績的取得，離不開 16 年來各位朋友和親友的支持與幫助。他們都是推動我們奮力前行的良師益友。希望讀者能夠從中獲得美的享受，為弘揚中國的旗袍文化，多作貢獻。

宋路霞 / 周鐵芝

2024 年 1 月 15 日

Preface

This book is the fourth qipao book of ours, following *A Rare Appreciation of the Qipaos Worn by the Grande Dames of Shanghai*, *A Collection of Qipaos from China's Prominent Families*, and *A Collection of Qipaos from China's Prominent Families (II)*.

This book contains 59 qipaos from 26 older generation ladies of Shanghai and Hong Kong collected by Shanghai Old Qipaos Collection Museum in recent years, including 2 qipaos from former Hong Kong Chief Executive Ms. Carrie Lam Cheng Yuet-ngor, 6 qipaos from Ms. Yung Liu Huixiu (Mr. Yung Tsung-chin's wife), 5 qipaos from Ms. Laura C. Soong (Mr. T.V. Soong's wife), 2 qipaos from Ms. He Baoshan (Mr. Qi Rushan's granddaughter), 3 qipaos from Dr. Delia Chen Chi Kuen Pei (a famous educator in Hong Kong), 7 qipaos from Ms. Fan Yuen Fong (a descendant of Tianyi Pavilion Library in Ningbo), 7 qipaos from Ms. Dora Tang Tai Yuet Wah (a famous children's educator in Hong Kong), and 2 qipaos from famous actress Lee Lai-wa ... Worn by the older generation ladies, all these qipaos feature a unique aesthetic concept of Chinese and Western clothing combined. They also witnessed the technical development in textile, printing, dyeing, design, cutting, sewing and embroidery at that time. Through an introduction to these qipaos, we try to show the essence of Shanghai-style qipaos in the past century. Like the three previously published qipao books, we will leave some real records on the development history of Shanghai-style qipaos.

Shanghai Old Qipaos Collection Museum was established 16 years ago. Ever since, with the support and assistance of our friends, we have traveled abroad to collect and rescue the qipaos worn by the older generation ladies, with the aim of leaving a rare, non-renewable and fleeting true record on Shanghai-style qipaos and a treasure trove of excellent qipao resources for future generations.

So far, we have collected over 600 qipaos from the wardrobes of the older generation ladies in New York, Los Angeles, San Francisco, Tokyo, Singapore, Hong Kong and Shanghai, and they have formed a family series, such as the Soong Ching-ling family, the Sheng Hsuan-huai family, the Yung Tsung-chin family, the New Shang-chow family, the V.K. Wellington Koo family and the Hsu Sze-hao family. There are many other exquisite qipaos worn by the older generation ladies of Shanghai and Hong Kong. Among them, 117 qipaos from Yung Tsung-chin's family have been donated to Shanghai University Museum, in order to leverage its advantages in exhibition, preservation and research to better play the inheritance role of these precious qipaos.

For the past 16 years, among all the older generation ladies who provided us with their qipaos or helped us receive qipaos, 39 have left us. We miss them so. Fortunately, although they have gone, their qipaos have returned to Shanghai. This makes us more aware of the importance and urgency of this "Old Qipaos Rescue Project". If we do not collect them from all over the world as soon as possible, Shanghai-style qipaos in their prime will become mere talk on paper due to the lack of physical objects, and the history of Chinese qipao culture will be missing a link. We are lucky to have personally participated in this very meaningful project with the assistance of these older generation ladies. The precious memories over the past 16 years have been left in our mind forever.

The qipaos contained in this book are only a small part of our "Old Qipaos Rescue Project". They are different in style, fabric and source. It is really fortunate to be able to gather them together in Shanghai.

No matter looking ahead to classics, paying tribute to classics, or studying classics, the ultimate goal is to inherit classics. In the appendix to this book, we present some of the famous qipaos replicated by Shanghai Zhi Lian Fu Culture Co., Ltd. We welcome comments from experts.

These achievements are inseparable from the support and assistance of our friends over the past 16 years. It is they who drive us forward. I hope that readers can enjoy beauty from this book and make more contributions to promoting Chinese qipao culture.

Song Lu-xia, Zhou Tie-zhi

Jan. 15, 2024

邊陳之娟・博士

Dr. Delia Chen Chi Kuen Pei

(Chen Zhi-juan)

香港著名的教育家、慈善家、太平紳士。她創辦的方方樂趣教育機構已有半個多世紀，是香港最早實行中英文雙語教育的大型綜合學校，如今已碩果累累，桃李滿天下。邊陳之娟博士心胸開闊，富有愛心，有高度的社會責任感，身兼多種社會職務，如中國少年兒童基金會理事、上海市歸國華僑聯合會理事、上海市政協委員（九屆、十屆）、香港兒童軍總會香港總監暨副會長、新華通訊社香港分社香港地區事務顧問……她百忙中仍不忘弘揚中國的旗袍文化，在重要場合總是一身漂亮的旗袍。本書刊出的這件橙色珠繡旗袍，是她 1963 年結婚時穿的禮服，胸部以上爲透明紗地，魚鱗刺繡間點綴透明水珠，橙色珠串形成了流蘇效果，使旗袍整體動感十足。另外一件是漳絨面料的長袖旗袍，還有一件是淺色真絲繡邊的無袖旗袍，均爲多年前的老旗袍。承蒙邊陳之娟博士厚愛，已捐贈給上海老旗袍珍品館。

邊陳之娟夫婦的結婚照（1963 年）

Dr. Delia Chen Chi Kuen Pei is a famous educator, philanthropist and Justice of the Peace in Hong Kong. The Fangfang Fun Educational Institution that she founded has been existing for over half a century. It was the first large-scale comprehensive school in Hong Kong to implement Chinese and English bilingual education. Throughout the years, it has achieved a great success and has students all over the world. Dr. Delia Chen Chi Kuen Pei is always with an open-minded view. She is a caring person and has an unwavering sense of social responsibility. She holds a variety of social positions, such as Director of China Children's Foundation; Director of Shanghai Federation of Returned Overseas Chinese, member of the CPPCC Shanghai Committee (Ninth and Tenth Sessions), Hong Kong Director and Vice President of Hong Kong Children's Army Association, and Advisor for Hong Kong Affairs of Xinhua News Agency Hong Kong Branch. She is also a Director and Vice President of the Scout Association of Hong Kong; a Hong Kong Regional Affairs Advisor of Xinhua News Agency Hong Kong Branch. She is a champion for promoting China's qipao culture despite her busy schedule, and always wears beautiful qipaos on important occasions. The orange bead-embroidered qipao displayed in this book is the dress she wore when she got married in 1963. It has a transparent gauze ground above the chest, with glittering droplet shaped beads embellished between the fish scale embroidery, and the orange beads forming a fringe effect, making the overall qipao dynamic. The other piece is a long-sleeved qipao made of Chinese traditional velvet. The third is a sleeveless one with light-colored silk embroidery. They are all old qipaos designed and tailored from many years ago. Thanks to the kindness of Dr. Delia Chen Chi Kuen Pei that they have been now donated to Shanghai Old Qipaos Collection Museum.

陳香梅·女士
Mrs. Anna Chan Chennault
(Chen Xiang-mei, 1925—2018)

祖籍廣東佛山，生於北京，是著名的華僑領袖、社會活動家、記者和作家。她在抗戰中擔任中央通訊社昆明分社記者，是當時中央社的第一位女記者。抗戰勝利後她在上海工作，並成為飛虎隊隊長陳納德將軍的夫人，此後一直活躍在政壇。她與陳納德將軍結婚時身穿一件大紅絲絨的珠繡旗袍，胸前各色珠粒在金色的盤帶中熠熠閃光，具有典型的中西合璧的海派風格，非常獨特。她有『中美民間大使』的美譽，數十年間致力於中外文化交流，擔任陳香梅教育基金會董事長，還是北京師範大學等多所院校的顧問和教授。本書刊出的這件旗袍是由上海芝蓮福文化有限公司複製的。

Born in Beijing, Mrs. Anna Chan Chennault (1925—2018) was a native of Foshan, Guangdong Province. She was a famous overseas Chinese leader, social activist, journalist and writer. During the War of Resistance Against Japanese Aggression, she served as a reporter for the Kunming Branch of the Central News Agency and was the first woman who worked in that capacity at the time. After the victory of the war, she worked in Shanghai and became the wife of General Chennault, the captain of the Flying Tigers. She had been active in politics ever since then. At the wedding ceremony she married General Chennault, she wore a bright red bead-embroidered velvet qipao, with various beads on the chest shining in the golden ribbon. It has a typical Shanghai-style fashion that combines Chinese and Western styles and is very unique. Mrs. Chennault had the reputation of "People's Ambassador between China and the United States". For decades, she had been committed herself to cultural exchanges between China and foreign countries. She had served as the chairman of Anna Chan Chennault Education Foundation and was also a consultant and professor at Beijing Normal University as well as many other universities. The qipao shown in this book is a replica of Mrs. Chennault's wedding dress made by Shanghai Zhi Lian Fu Culture Co., Ltd.

陳天眞・女士
Ms. Chen Tianzhen
（Chen Tian-zhen, 1928—2008）

上海百樂門創辦人顧聯承先生的兒媳、顧森康先生的夫人。她是江蘇人，從小在上海長大，入讀聖約翰大學，有校花之譽。她從事過會計職業，還是經營股票的好手。1949年與丈夫定居香港。她終生喜歡穿旗袍。她的旗袍不僅數量多，而且面料考究，花色繁多，做工精細，面料都是真絲或毛料的。2018年，她的兒子顧家璉先生將她的15件美麗旗袍從香港和加拿大帶到上海，捐贈給上海老旗袍珍品館。本書收錄她的三件旗袍，均爲真絲印花，非常靚麗。

顧森康、陳天真夫婦

Ms. Chen Tianzhen (1928–2008) was the daughter-in-law of Mr. Gu Liancheng, who was the founder of Shanghai Paramount, and the wife of Mr. Gu Senkang. Ms. Chen was a native of Jiangsu Province but grew up in Shanghai. She attended St. John's University in Shanghai and at the time was known as the campus beauty. She later had established a career in accounting field and was also successful at trading stock. Since 1949, Ms. Chen and her husband had been living in Hong Kong. She loved wearing qipaos all the time. She had numerous qipaos, which are made of exquisite fabrics, with various colors and fine workmanship, either by silk or by wool. In 2018, her son Mr. Gu Jialian brought 15 of her beautiful qipaos to Shanghai from Hong Kong and Canada, and donated them to Shanghai Old Qipaos Collection Museum. This book contains three of her qipaos, all of which are silk-printed and exquisitely beautiful.

杜美霞・女士
Ms. Du Meixia
(Du Mei-xia, 1931—2019)

上海高橋人，青幫大亨、海上聞人杜月笙次女，民國著名女老生孟小冬的義女，浙東商業銀行董事長、黃金大戲院經理金廷蓀的兒媳。她常年伴隨、照顧『冬皇』孟小冬，致力於京劇推廣，曾擔任孟小冬女士國劇獎學基金會董事長。這件絲絨旗袍套裝是杜美霞晚年訂製的，參加京劇票房和公眾活動時常穿，後贈送給臺灣程派名票陸寒波女士。陸寒波女士晚年病重時，連同自己的京劇戲服一起交由臺灣名伶、乾旦大家夏華達收藏。夏華達先生九十九歲往生前，輾轉由臺北帶回上海，交由徒侄吳崢嶸先生保存，後由吳崢嶸先生捐贈給上海老旗袍珍品館。

Ms. Du Meixia (1931–2019) was born in Gaoqiao, Shanghai. She was the second daughter of Du Yuesheng, a tycoon of Qing Gang in Shanghai, and the adopted daughter of Meng Xiaodong, a famous female actress playing old-man role in Peking Opera. She was also the daughter-in-law of Mr. Jin Tingsun, the chairman of East Zhejiang Commercial Bank and manager of the Golden Theater. For a long period of time, Ms. Du was so committed to promoting Peking Opera that she had been accompanying and taking care of the so-called "Dong Empress" Meng Xiaodong all year round. She once served as the chairman of Ms. Meng Xiaodong Chinese Opera Scholarship Foundation. The velvet qipao suit shown in this book was customer-made as ordered by Ms. Du Meixia in her later years. She often wore it when attending amateur Peking Opera activities or important public events. She later gave it to Ms. Lu Hanbo, a famous Taiwanese Cheng Yanqiu genre amateur Peking Opera actress. When Ms. Lu Hanbo became seriously ill in her later years, she gave this suit together with her Peking Opera costumes to Mr. Xia Huada, a famous Taiwanese actor in Peking Opera playing woman-role. Before Mr. Xia Huada passed away at the age of ninety-nine, the qipao suit along with the costumes were brought back to Shanghai from Taipei and handed over to his apprentice and nephew Mr. Wu Zhengrong for preservation. Mr. Wu Zhengrong later donated it to Shanghai Old Qipaos Collection Museum.

杜美霞（右）與義母孟小冬（左坐者）合影

鄧戴月華·女士

Ms. Dora Tang Tai Yuet Wah

(Dai Yue-hua, 1951—)

祖籍廣東番禺，出生在一個祖輩讀書做官的狀元家族。其父戴文輝先生是香港文華書局的創辦人，也是一位業餘書法家和藝術家。她1971年赴美國留學，學費全免，是紐約州立大學醫療技術專業學士，又獲獎學金讀研究生，獲紐約州首府醫學院免疫學碩士學位。回香港後先是從事醫務工作，1990年開始從事幼兒早期智力開發教育，用她自創的自然反射方法及特殊的語音系統去引導幼兒讀寫英語，取得了行之有效、立竿見影的效果，深受幼兒家長歡迎，成爲香港兒童啓蒙教育專家，DOORS Method learning Centre 創辦人。她從1981年開始穿旗袍。她的旗袍無論是面料還是款式，都很精美，是上海師傅的精湛手藝。本書收錄的七件旗袍：藍色金彩提花喬其紗短袖旗袍，綠色格紋提花印花無袖旗袍，白地金彩提花喬其紗短袖旗袍，寶藍色樹葉紋短袖旗袍，黑色蕾絲珠繡短袖旗袍，黑、紅雙色蕾絲盤帶繡中袖旗袍，玫紅蕾絲珠片繡長袖旗袍。七件旗袍的盤扣多爲填芯花扣，非常精緻。

Ms. Dora Tang Tai Yuet Wah or Madame Tang (1951-) is a native of Panyu, Guangdong Province. Her ancestors reportedly were highly accomplished scholars sent from north China as government officials. Her father Mr. Tai Man Fai is an amateur calligraphist and artist as well as the founder of the Man Wah Bookstore in Hong Kong. In 1971, Madame Tang was awarded a full scholarship for her study in Medical Technology at the State University of New York at Albany, and an additional fellowship for her graduate study in Immunology at Albany Medical School. After returning to Hong Kong, Madame Tang did some medical work. In 1990, she began to engage in early childhood intellectual development education. She researched, designed, and developed a unique phonics system to teach English to children at toddlers' age, using a reflexive approach called the "DOORS Method". Its overwhelming reception eventually led to the founding of her school DOORS Method Learning Centre. In 1981, Madame Tang began wearing qipaos. Whether it is the fabric, the style, or the colour, her qipaos are all individually, meticulously, and exclusively hand-sewn by Shanghainese tailor masters. In this book, seven qipaos from Madame Tang's collection are presented: a blue georgette short-sleeved qipao with golden jacquard patterns, a green jacquard and printed sleeveless qipao with checkered patterns, a white georgette short-sleeved qipao with golden jacquard patterns, a sapphire blue short-sleeved qipao with leaf patterns, a black lace bead-embroidered short-sleeved qipao, a black-red lace ribbon-embroidered mid-sleeved qipao, and a rose red lace bead-embroidered long-sleeved qipao. The buttons of the seven qipaos are mostly filled ones, which is very exquisite.

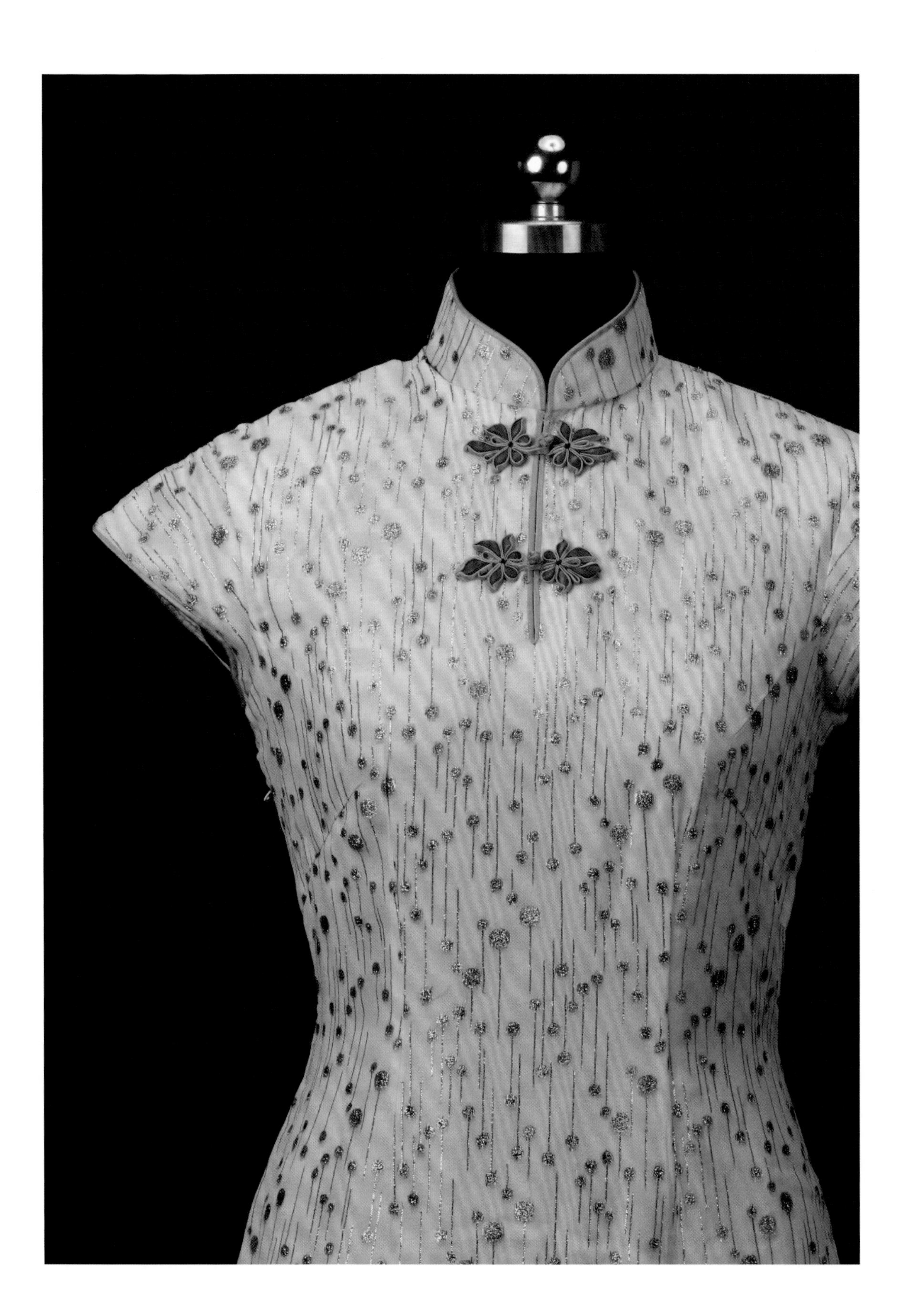

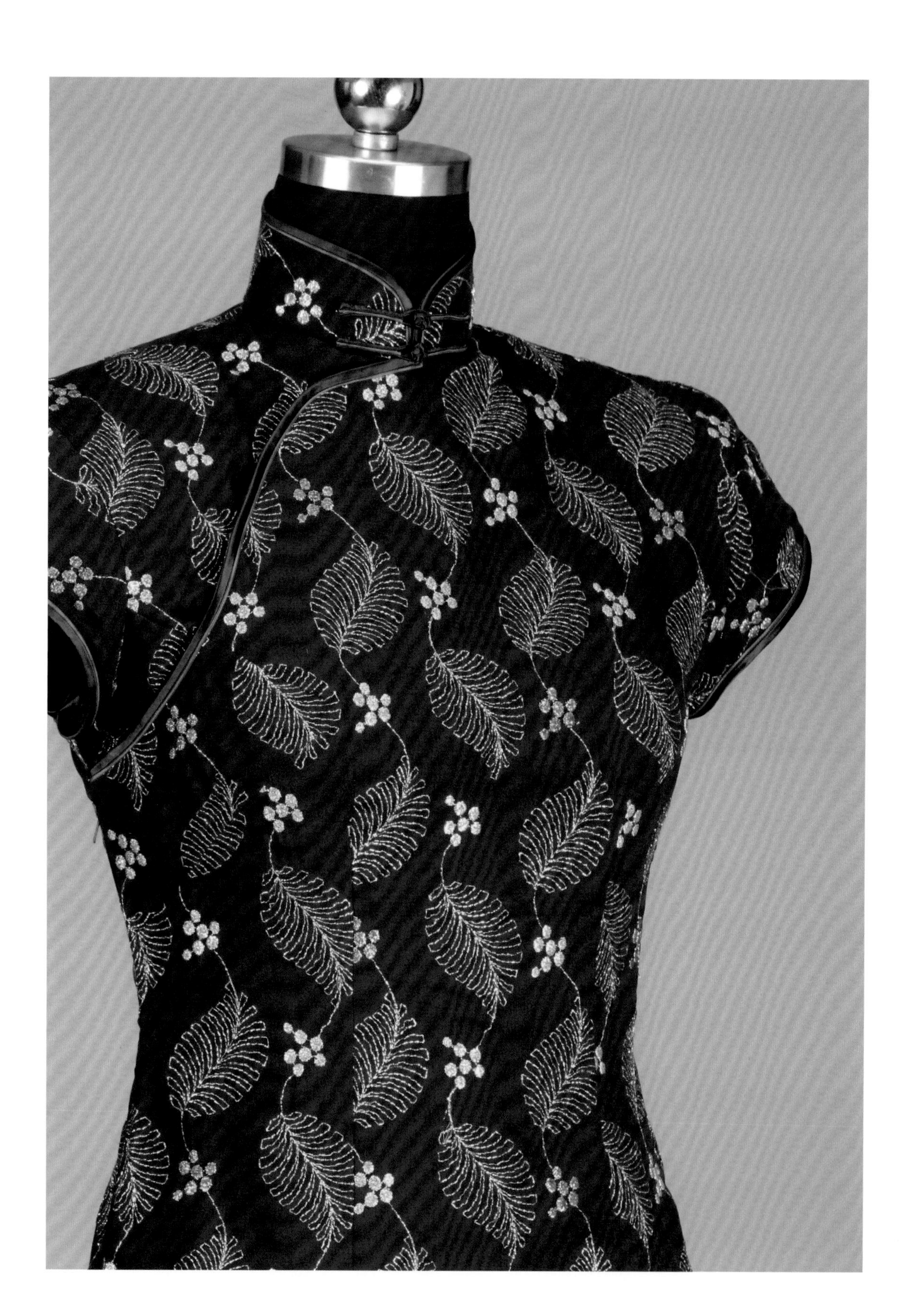

范元芳·女士
Ms. Fan Yuen Fong
（Fan Yuan-fang, 1948— ）

1948年出生在上海，16歲與家人移居香港，在她的忘年交陳天真女士（上海百樂門創辦人顧聯承先生的兒媳）的影響下，穿了半個多世紀的旗袍，形成了自己獨特的旗袍風格，是香港最會穿旗袍的典範之一。本書收錄了她7件旗袍，其中5件是旗袍套裝，均是她在各種社交場合的禮服，色彩鮮豔，面料考究，製作精良。粉色與綠色的旗袍套裝款式相似，外套使用對襟挖領設計，皆在蕾絲面料上做盤帶繡和貼花裝飾；一件黑地絲絨套裝，外套採用大花絲絨面料，平添了雍容華貴的亮色；另一件黑色旗袍套裝，外套是著名品牌聖約翰的產品；還有一件是紅地藍花的真絲旗袍套裝。其餘兩件旗袍，一紅一綠，均是真絲面料，有精美的三條緄邊，做工非常精細。這些旗袍，承蒙范元芳女士厚愛，捐贈給上海老旗袍珍品館永久珍藏。

Ms. Fan Yuen Fong was born in Shanghai in 1948. She moved to Hong Kong with her family at the age of 16. Under the influence of her old friend Ms. Chen Tianzhen (the daughter-in-law of Mr. Gu Liancheng, the founder of Shanghai Paramount), she has been wearing qipaos for more than half a century and has developed her own unique style of wearing it. She is one of the best qipao-wearing models in Hong Kong. Among seven qipaos of hers displayed in this book, five are qipao suits that are all dresses Ms. Fan Yuen Fong wore on various social occasions. They are brightly colored and well-made with exquisite fabrics. The pink and green qipao suits are similar in style. Both suits have a coat that is designed as Chinese-style opening, and are decorated with ribbon embroidery and appliqués on the lace fabric. One qipao suit is made of black velvet fabric, with the coat made of large floral velvet fabric, adding a graceful and luxurious style; another is a black qipao suit, with the coat of the famous St. John's brand; the third is a silk qipao suit with blue flowers on a red background. Among the remaining two qipaos, one is red and the other is green. Both are made of silk fabric and have exquisite three piping edges with an excellent workmanship. Thanks to Ms. Fan Yuen Fong, these qipaos were donated to Shanghai Old Qipaos Collection Museum as a permanent collection.

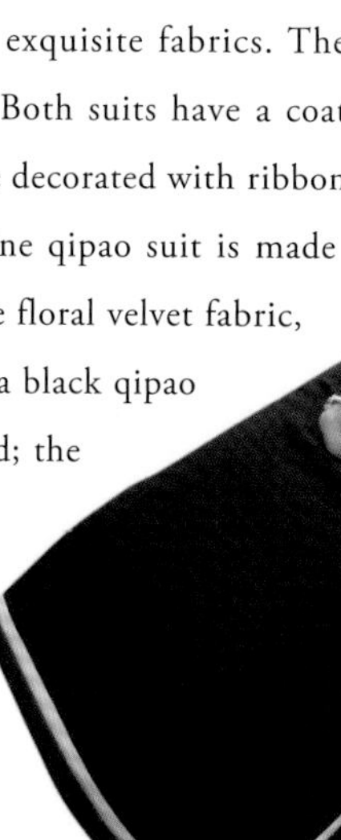

賀寶善・女士
Ms. He Baoshan

（He Bao-shan, 1928—2015）

祖籍湖南長沙，生於北京，從小在外公齊如山先生身邊長大。受家學浸潤，她於中國傳統文化頗有根底，於音樂、書畫、寫作、翻譯、京劇均有造詣。她在北京燕京大學音樂系畢業後，曾教授鋼琴與中文，數十年間學貫中西，有才女之譽，1951年與太古洋行買辦姚剛先生在香港結爲伉儷，六十餘年來甘苦與共，相夫教子，堪稱楷模。近年來有《思齊閣紀事》《思齊閣憶舊》問世。她的這件米色珠繡旗袍套裝，領面和袖口均繡以細長的珠串，珠光寶氣，璀璨華麗，似珊瑚又似花卉，令人耳目一新。另一件也是米色的旗袍套裝，其外套的領子用重疊的葉瓣裝飾，非常罕見。這兩件旗袍均由她的女兒姚詠蓓女士提供，經香港的吳順昇先生聯絡，捐贈給上海老旗袍珍品館，由宋路霞、曹蘭珍帶回上海。

Ms. He Baoshan (1928–2015), a native of Changsha, Hu'nan Province, was born in Beijing. She grew up with her maternal grandfather, Mr. Qi Rushan. Influenced by her family, she had established a solid foundation in traditional Chinese culture and had attainments in music, Chinese calligraphy and painting, writing, translation, and Peking Opera. After graduating from the music department of Yenching University in Beijing, she taught piano and Chinese. Over the decades, she had been known as a talented woman for her rich knowledge in both Chinese and Western culture. In 1951, she married Mr. Yao Gang, a comprador of Swire Pacific, in Hong Kong. The couple had been enjoying life and enduring hardships for more than sixty years. Sharing weal and woe, caring for husband and raising children, Ms. He can be called a role model. In recent years, her works *Chronicles of Siqi Pavilion* and *Remembering the Past of Siqi Pavilion* have been published. Two of her qipaos are shown in this book. One is a beige bead-embroidered qipao suit with slender beads embroidered on the collar and cuffs. Bright and gorgeous like jewels, corals and flowers, its style is refreshing. The other is also a beige qipao suit. The collar of the coat is decorated with overlapping leaves, which is very rare. Both qipaos were provided by her daughter Ms. Yao Yongbei. After contacted by Mr. Otto Wu from Hong Kong, they were brought back to Shanghai by Ms. Song Lu-xia and Ms. Cao Lanzhen and donated to Shanghai Old Qipaos Collection Museum.

林鄭月娥・女士
Ms. Carrie Lam Cheng Yuet-ngor
（Zheng Yue-e, 1957— ）

祖籍浙江舟山，1957年出生在香港，是香港金紫荊星章、香港大紫荊勳章獲得者、香港特別行政區第五任行政長官。她1980年香港大學畢業後進入香港政府，2000年後曾任香港特區政府社會福利署署長、發展局局長及政務司司長。她『帶領香港特區政府勇於擔當，積極作爲，堅定維護「一國兩制」的方針和基本法，認真謀劃香港的長遠發展，積極參與粵港澳大灣區發展和落實「一帶一路」，著力破解事關廣大居民切身利益的問題，努力爲青年人成長和發展創造條件，體現了「志不求易」「事不避難」的精神，取得了良好成績。中央對林鄭月娥行政長官和特別行政區政府的工作是充分肯定的。』（國家主席習近平語）。她非常喜歡旗袍，在重要場合總是一身精美的旗袍。她的旗袍套裝尤其典雅、端莊，堪稱職業女性服飾的典範。本書收錄的兩件旗袍均是在蕾絲網紗機繡面料上加手工珠繡，一件是淺灰色無袖長旗袍，另一件是灰色短袖短旗袍，非常精緻，令人過目難忘。

Ms. Carrie Lam Cheng Yuet-ngor, a native of Zhoushan, Zhejiang Province, was born in Hong Kong in 1957. She is a recipient of Hong Kong Gold Bauhinia Star and Hong Kong Grand Bauhinia Medal, and is the fifth-term Chief Executive of Hong Kong Special Administrative Region. She entered the Hong Kong government after graduating from the University of Hong Kong in 1980. After 2000, she served as Director of Social Welfare, Secretary for Development and Chief Secretary of Administration of the Hong Kong SAR Government. She "has led the Hong Kong SAR Government to be courageous and active in its responsibilities, firmly upholding the policy of 'one country, two systems' and the Basic Law. She has conscientiously planned for the long-term development of Hong Kong, and actively participated in the development of Guangdong-Hong Kong-Macao Greater Bay Area and the implementation of the 'Belt and Road Initiative'. She has striven to solve problems that concern the vital interests of the vast majority of people, and to create conditions for the growth and development of young people. Ms. Carrie Lam Cheng Yuet-ngor has embodied the spirit of 'never seek what is easy', thus achieving good results. The central government fully endorses the work of Chief Executive Ms. Carrie Lam Cheng Yuet-ngor and the Hong Kong SAR Government." (President Xi Jinping's words). She likes qipaos very much and always wears an exquisite qipao on important occasions. Her qipao suit is particularly elegant and dignified, and can be regarded as a model of professional women's clothing. The two qipaos shown in this book are made of machine-embroidered lace with hand-made bead embroidery. One is a light gray sleeveless long qipao, and the other is a gray short-sleeved one. They are very exquisite and will leave people an unforgettable impression.

劉福曾・女士

Ms. Liu Fuzeng

(Liu Fu-zeng, 1918—2002)

祖籍安徽盧江，生在上海。她是晚清淮軍名將劉秉璋的曾孫女，又是李鴻章的曾外孫女（其母親李國華是李鴻章的長孫女），1939年畢業於上海著名的中西女中，1943年畢業于上海聖約翰大學，後在大陸銀行工作，解放後任市東中學數學和英文教師，直至退休。她的這件粉色薄呢長袖旗袍，設計簡單，傳統式樣，不作修飾，長期由她的兒子兒媳葉世楝、李麗夫婦保存，前年捐贈給上海老旗袍珍品館。

中年劉福曾與兒子葉世楝

Ms. Liu Fuzeng (1918—2002), a native of Lujiang, Anhui Province, was born in Shanghai. She was the great-granddaughter of Liu Bingzhang, a famous Huai Army general in the late Qing Dynasty, and the maternal great-granddaughter of Li Hongzhang (her mother, Li Guohua, is Li Hongzhang's eldest granddaughter), a famous statesman in the late Qing Dynasty. She graduated from the famous McTyeire School for Girls in Shanghai in 1939, and then graduated from Shanghai St. John's University in 1943. Later, she worked in Continental Bank and then after 1949, she worked as a mathematics and English teacher at Shanghai East City Middle School until retirement. The pink thin woolen long-sleeved qipao shown in this book has a combination of simple design and traditional style, but quite modest. This qipao had been preserved by her son Ye Shidong and daughter-in-law Li Li for a long time, and was donated to Shanghai Old Qipaos Collection Museum two years ago.

劉蓮華・女士
Ms. Rio L. Chiang
（Liu Lian-hua, 1930— ）

上海灘著名實業家劉吉生先生的小女兒、劉鴻生先生的侄女。她出生在巨鹿路675號美麗的愛神花園裡（現爲上海作家協會機關所在地），母親陳定貞是南潯望族之後。她從小受到良好的教育，先後在中西附小和教會女中畢業，1948年留學美國，並在美國結婚成家，成爲丈夫的賢內助。受母親影響，她年輕時就擅長女紅，無論是編織、裁剪、設計、縫紉均駕輕就熟，尤其對於自己的旗袍，她總是樂於親自設計，很講究面料和色彩，在款式和搭配上常有中西合璧的獨到之處。現在她已高齡，仍樂於縫紉和編織，是兒女和孫輩們服飾打扮的最佳顧問。她的這件灰色珠繡旗袍套裝，外套設計非常特別，類似連衫裙，由徐景燦、宋路霞從美國帶回上海。

Ms. Rio L. Chiang (1930-), the youngest daughter of Liu Jisheng and the niece of Liu Hongsheng, both were famous industrialists of Shanghai, was born in the beautiful Cupid Garden at No. 675 Julu Road, Shanghai (now the seat of Shanghai Writers' Association). Her mother, Chen Dingzhen, was a descendant of a prominent Nanxun family. With such a family background, Ms. Rio L. Chiang received a good education from her childhood on. She graduated from Primary School Affiliated to McTyeire School for Girls and a missionary middle school for girls. In 1948, Ms. Rio L. Chiang went to the United States to study and later got married there. She then became her husband's virtuous helper. Inspired by her mother, Ms. Rio L. Chiang is good at needlework ever since her youthful days, including weaving, cutting, designing and sewing. Especially for her own qipao dress, she always likes to design it by herself. She is very selective about fabrics and colors, and often has her own unique styles and matching. Her design shows her special ability in combining Chinese and Western elements. Now at an advanced age, she still enjoys sewing and knitting, and is the best consultant for her children and grandchildren. This gray bead-embroidered qipao suit of Ms. Rio L. Chiang's, with a very exquisite coat that resembles a full-length dress, was brought back to Shanghai from the United States by Ms. Jeanette Zee and Ms. Song Lu-xia.

青年劉蓮華（右）與席與時

李麗華・女士

Ms. Lee Lai-wa

(Li Li-hua, 1924—2017)

爲我國影壇巨星，祖籍河北，出生在上海，父母都是我國著名京劇演員。她16歲步入影壇，1940年因主演電影《三笑》一舉成名，此後共主演了120餘部影片，2015年獲臺灣第52屆金馬終身成就獎。她身材苗條，顧盼有儀，一生與旗袍結下不解之緣。她的這件玫瑰大花的短袖長旗袍，構圖誇張，色彩豔麗，大紅大紫，是其晚年穿的旗袍，由其女兒提供，經香港的吳順昇先生聯絡，連同其他3件旗袍均捐贈給上海老旗袍珍品館，分別由徐景燦、宋路霞、舒康健從香港帶回上海。

Ms. Lee Lai-wa (1924–2017) was a superstar in Chinese film industry. She was a native of Hebei Province, but was born in Shanghai. Her parents were famous Peking Opera artists. Ms. Lee entered the film industry at the age of 16 and quickly became famous in 1940 for starring in the movie *Three Smiles*. Since then, she had starred in more than 120 films. In 2015, she won the 52nd Golden Horse's Lifetime Achievement Award in Taiwan. She was slender and elegant, and had been fond of wearing qipaos throughout her life. This short-sleeved long qipao with large roses has an extravagant composition and bright colors of red and purple. It is the qipao she wore in her later years, and was provided by her daughter, after contacted by Mr. Otto Wu in Hong Kong. It was donated to Shanghai Old Qipaos Collection Museum together with the other three qipaos which were brought back to Shanghai from Hong Kong by Ms. Jeanette Zee, Ms. Song Luxia and Mr. Shu Kangjian respectively.

龐左玉・女士
Ms. Pang Zuoyu
(Pang Zuo-yu, 1915—1969)

浙江湖州南潯人，又名龐昭，别署瑤草廬主，著名畫家，是南潯『四象』之一龐氏家族後裔，近代著名收藏大家龐元濟的侄女，海上畫壇『四公子』之一樊伯炎先生的夫人。她1934年畢業於上海藝術專科學校，後從鄭曼青先生學習花卉。由學徐青藤、陳白陽入手，兼得其伯父龐元濟指導，廣臨古畫，畫藝大進。她精於花卉、蟲草，風神秀雅，筆致工穩，民國時期曾參加中國女子書畫會、中國畫會，系『清遠藝社』成員。建國初，上海籌建中國畫院，其爲第一批進入畫院的九位女畫家之一。或許由於她的名字中有個『左』字，所以她的旗袍也從左邊開襟。這件綠色真絲提花旗袍，是她留下來的唯一一件旗袍，由青年收藏家吳崢嶸先生捐贈給上海老旗袍珍品館。

Ms. Pang Zuoyu (1915–1969) was born in Nanxun, Huzhou, Zhejiang Province, also known as Pang Zhao. She was a famous Chinese painter and the master of Yao Cao Lu, a private art studio. She was a descendant of the Pang family, one of the so-called "Four Elephants", a prominent clan in Nanxun. Ms. Pang was the niece of Mr. Pang Yuanji, a famous collector in modern times, and the wife of Mr. Fan Boyan, one of the famous "Four Young Painters" in Shanghai. Ms. Pang graduated from Shanghai Art College in 1934 and later studied Chinese flowers painting from a master, Mr. Zheng Manqing, starting from studying the styles and techniques of Xu Qingteng and Chen Baiyang, both were Chinese classical painting masters. At the same time, she was also receiving the guidance of her uncle Pang Yuanji. Since she extensively studied ancient paintings and her painting skills improved greatly. With a graceful style and steady workmanship, she was good at painting flowers, insects and plants. During the Republic of China, she participated in the Chinese Women's Calligraphy and Painting Association and the Chinese Painting Association, and was a member of "Qingyuan Art Society". As Shanghai prepared to establish a Chinese Painting Academy during the early days of New China, she was one of the first nine female painters to enter the academy. Perhaps because her name contains a Chinese character meaning "left" in English, her qipaos open from the left. This green silk jacquard qipao is the only one left by her. It was donated to Shanghai Old Qipaos Collection Museum by young collector Mr. Wu Zhengrong.

任蕊芬・女士

Ms. Ren Ruifen

(Ren Rui-fen, 1930—)

江蘇宜興人，從小生活在上海，在常熟路安福路的一棟漂亮的花園洋房裡居住多年。她的曾祖父任道鎔是晚清山東巡撫（省長）、浙江巡撫、河道總督，祖父與父親都是那個時代讀書做官的典型。她從小受家族傳統文化的薰陶，爲人厚道，性情豁達，擅長烹飪，婚後相夫教子、勤儉持家。她與丈夫盛毓綬先生（晚清洋務巨擘盛宣懷的孫子）1977年移居香港，曾多年擔任在港日本人和香港人的正宗上海菜教師。旗袍是她日常服裝中的最愛。這件旗袍使用爛花絲絨面料製作，爲深綠色花卉紋樣，邊緣使用同色系緄邊，偏襟處綴有精美的蝴蝶盤扣，十分生動，是由她的女兒盛承慧女士提供的。

盛毓綬、任蕊芬夫婦青年時代

Ms. Ren Ruifen (1930-) is a native of Yixing, Jiangsu Province. She has been living in Shanghai since she was a child and lived in a beautiful garden house on Anfu Road at the corner of Changshu Road for many years. Her great-grandfather Ren Daorong was the governor of Shandong Province and Zhejiang Province respectively, and also a governor of the river authority in the late Qing Dynasty. Ms. Ren's grandfather and father were both typical scholars who later became officials in that era. Influenced by her family's tradition, Ms. Ren is kind and open-minded. She is good at cooking, takes care of her husband and children, and is diligent and thrifty in running the house. She and her husband, Mr. Sheng Yushou (the grandson of Sheng Hsuanhuai, the late Qing Dynasty's Westernization tycoon) moved to Hong Kong in 1977, there She worked as a teacher of authentic Shanghainese cuisine for Japanese and Hong Kong residents for many years. Qipao is her favorite among daily wear. The qipao shown here is made of burnt-out velvet fabric with a dark green floral pattern. The edges are piping in the same color and the opening is embellished with exquisite butterfly buttons. This qipao was provided by her daughter, Ms. Sheng Chenghui.

榮劉惠秀・女士

Ms. Yung Liu Huixiu

(Liu Hui-xiu, 1900—1994)

榮劉惠秀女士60歲生日時與兒子榮鴻慶先生合影

江蘇無錫人，是我國著名實業家、『麵粉大王』和『棉紗大王』榮宗敬先生的夫人。老人家性情溫和，知書達理，教子有方。1938年榮宗敬先生在香港不幸逝世時，他們的兒子榮鴻慶才15歲，女兒榮卓如17歲，在她無微不至的照料和培育下，兒女均茁壯成長。榮鴻慶先生經過多年不懈奮鬥，成爲著名的銀行家和實業家；榮卓如女士是大陸改革開放後，率先投資家鄉紡織工業的先驅。老人家留下了6件精美旗袍，分別是：藍黃提花有立體浮雕效果的旗袍套裝；淺藍色真絲提花的中袖旗袍；米色機繡蕾絲面料的旗袍套裝；天藍色真絲印花的旗袍套裝；淺綠色起縐肌理的旗袍套裝；紅色真絲印花的旗袍套裝。這些旗袍大多採用高檔進口面料，色澤秀雅，做工精良，非常珍貴，由榮鴻慶先生的夫人榮周淑霞女士提供。

Ms. Yung Liu Huixiu (1900–1994) was born in Wuxi, Jiangsu Province. She was the wife of Mr. Yung Tsung-chin, a famous industrialist in China who had the nicknames of the "King of Flour" and the "King of Cotton Yarn". Ms. Yung was well-educated, with a kind temperament and a good sense of judgement. When her husband unfortunately passed away in Hong Kong in 1938, their son Yung Hongqing was only 15 years old, and their daughter Lily Y. Hardoon, merely 17. Under Ms. Yung's meticulous care and nurturing, both children thrived. After years of unremitting efforts, her son, Mr. Yung Hongqing, became a famous banker and industrialist; while her daughter Lily Y. Hardoon was one of the first investing in the textile industry in her hometown after mainland China adopted the reform and open-up policy. Ms. Yung left behind six exquisite qipaos, namely, a qipao suit made of light blue and yellow miscellaneous jacquard and three-dimensional relief fabric, a mid-sleeve qipao suit made of light blue silk jacquard, a qipao suit made of beige mesh machine-embroid fabric, a light blue silk-printed qipao suit, a light green silk jacquard qipao suit, and a red silk-printed qipao suit. Most of these qipao suits are made of high-end imported fabrics, with elegant colors and excellent workmanship. They are very precious and are provided by Ms. Yung Zhou Shuxia, the wife of Mr. Yung Hongqing.

施蓓芳・女士

Ms. Shi Beifang

（Shi Bei-fang, 1926—2021）

祖籍江蘇吳江，生在上海，是著名外交官施肇基先生的侄孫女、著名文史專家、上海文史館館員、詩人周退密先生的夫人。她從小跟姑祖母施彤昭生活在一起，住在青海路 44 號那棟大宅裡，熟知幾乎所有上海灘的豪門舊事，被譽爲上海大宅門的活字典。1949 年後她一直在中國工商銀行上海徐匯區分行工作，擔任業務指導，業內稱其爲銀行業的資深老法師。她從前一直穿旗袍。這件白底藍花的絲絨旗袍，是深臥她的箱底 60 年的老旗袍，由她本人親自翻找出來，捐獻給上海老旗袍珍品館。

Ms. Shi Beifang (1926–2021), a native of Wujiang, Jiangsu Province, was born in Shanghai. She was the grand-niece of the famous Chinese diplomat Mr. Shi Zhaoji, and the wife of Mr. Zhou Tuimi who was a famous literary and historical expert, a poet, and a librarian of Shanghai Museum of Literature and History. Since her childhood, Ms. Shi had lived with her great-aunt Shi Tongzhao in the mansion at No. 44 Qinghai Road, Shanghai. With a rich social background, she was familiar with almost all the old stories of Shanghai's wealthy families, and was known as the "walking dictionary" in this field. After 1949, she had been working with Shanghai Xuhui District Branch of the Industrial and Commercial Bank of China as a business director. In the banking industry, people called her "Senior Mage" for her extensive knowledge in the business. Ms. Shi used to wear qipaos all the time. This qipao is decorated with blue patterns on the white ground. It's an old one that had been lying at the bottom of her trunk for 60 years. She dug it out and donated it to Shanghai Old Qipaos Collection Museum.

宋胡靜君·女士
Ms. Soong Hu Jingjun
(Hu Jing-jun, 1923–2023)

祖籍江西九江，出生在九江，曾任小學音樂教師，1953年移居香港，是香港美孚石油公司總裁宋啓鄖先生的夫人。她美麗而長壽，享年100歲，青年時代就鍾情於旗袍，在海外生活了70多年，始終是一身漂亮的旗袍，而不穿西服，直到98歲高齡時，出門仍舊要仔細化妝，然後穿上華麗的旗袍，被譽爲百歲旗袍美人。她的旗袍不僅數量多，而且每一件都是由她自己精心設計，用料和做工都非常講究，款式也有別於傳統旗袍，多數採用半透明的蕾絲網紗繡花面料，裡面相應地加上襯裡。本書收錄她5件旗袍，分別是：黃地彩色蕾絲珠繡旗袍；紫色蕾絲珠繡旗袍；黑地蕾絲彩色貼花旗袍；咖啡色蕾絲珠片繡旗袍；草綠色織金紗旗袍。由她的女兒簡宋麗迪女士提供。

Ms Soong Hu Jingjun (1923–2023), born in Jiujiang, Jiangxi Province, grew up in Shanghai where she taught singing in a primary school. She married John L. Soong in Shanghai and moved to Hong Kong in 1953. After her husband retired, they moved overseas and spent their remaining years there. Being the wife of the Managing Director of Mobil Oil Company, her beauty and flair for fashion was much admired in Hong Kong society and she only wore qipaos, sometimes with a matching coat but mostly on its own. Her qipaos were renowned for their elegance and intricacies both in workmanship and fabrics. Never one to be traditional, Ms Soong designed her qipaos in keeping with modern trends and latest fashion. She lived until shortly after her 100th birthday and she still wore qipaos to parties into her 90s with makeup perfectly applied and hair beautifully styled. This book contains five pieces from her vast wardrobe, namely, a yellow-background colored lace bead-embroidered qipao, a purple lace bead-embroidered qipao, a black lace colored applique qipao, a coffee-colored lace sequin-embroidered qipao, and a grass-green golden-thread-woven mesh-like qipao. They were kindly loaned to us by her daughter Lydia Soong Kan.

王韻梅・女士
Ms. Wang Yunmei
(Wang Yun-mei, 1925—2002)

浙江紹興人，早年生活在上海，在1946年上海各界爲蘇北賑災而發起的『上海小姐』選美活動中獲得冠軍，成爲新聞人物，又因報章關於『幕後金主』插手的報導，而成爲更火的新聞人物。她是位藝術愛好者，愛好繪畫和攝影，開過照相館，與張大千、唐雲等藝壇大師相過從，1950去香港後，師從廣東籍的藝術家周公理，在油畫與水墨畫上下過苦功。她的丈夫是商界大亨王曉籟的兒子王守理，夫妻和睦，相敬如賓，兒孫繞膝，在美國度過人生的最後歲月。她的這件大紅緞面的繡花旗袍，是由她的外孫女王麗瑞女士特地從美國寄到上海的。

Ms. Wang Yunmei (1925—2002) was a native of Shaoxing, Zhejiang Province. She lived in Shanghai in her early years. In 1946, Ms. Wang won the championship in the "Miss Shanghai" beauty pageant organized by all walks of life in Shanghai for disaster relief in northern Jiangsu Province and became a media celebrity. Later, due to further news reports about "behind-the-scene money man's intervention", she became even more a media attraction. Ms. Wang was an art lover who liked painting and photography. She had opened a photo studio and studied with art masters such as Zhang Daqian and Tang Yun. After going to Hong Kong in 1950, she studied under the guidance of the Guangdong artist Zhou Gongli and worked very hard on practicing oil painting and Chinese ink painting. Her husband Wang Shouli was the son of Shanghai business tycoon Wang Xiaolai. The couple lived together in a harmonious and mutual respect way while always having children and grandchildren around them. They spent the last years of their lives in the United States. This bright red satin embroidered qipao was specially sent to Shanghai from the United States by her granddaughter Miss Lily Falzon Wong.

汪紫莘・女士

Ms. Wang Zixin

（Wang Zi-xin, 1897—1971）

祖籍安徽旌德，青少年時代在上海教會女校讀書，畢業於清心女中，精通英文和法文，擅長鋼琴演奏。她的丈夫朱有卿（耀廷，1893-1936）先生是美國康奈爾大學土木工程專業的高才生，與茅以昇先生是同專業的同學，學成回國後擔任浙江公路局局長，主持修建了首條從上海到杭州的公路。由於他政治上反蔣反日，不幸於1936年遭遇暗殺。丈夫犧牲後，汪紫莘獨自撫養5個孩子，含辛茹苦，把他們都培養成學有專長的技術人員。這件織錦緞面的狐狸皮旗袍，質地上乘，做工精細，非常罕見，由她的兒子兒媳朱曾滸、高麗霞夫婦捐贈給上海老旗袍珍品館。

Ms. Wang Zixin (1897−1971) was originally from Jingde, Anhui Province. She studied at a missionary girls' school in Shanghai and graduated from Pure Heart School for Chinese Girls. She was proficient in English and French and was good at piano performance. Her husband, Mr. Zhu Youqing (a.k.a. Yao Ting, 1893−1936), was a talented student majoring in civil engineering at Cornell University in the United States. He was a classmate of the same major as Mr. Mao Yisheng, a renowned bridge-building engineer in China. After graduation from Cornell University, Mr. Zhu served as the director of Zhejiang Highway Bureau and presided over the construction of the first highway from Shanghai to Hangzhou. Due to his political opposition to Chiang Kai-shek and Japanese invaders, he was unfortunately assassinated in 1936. After her husband's death, Ms. Wang Zixin raised their five children on her own, enduring hardships and training them all into skilled technicians. This brocade-covered fox fur qipao is of high quality and fine workmanship, and is very rare. It was donated to Shanghai Old Qipaos Collection Museum by her son Zhu Zenghu and daughter-in-law Gao Lixia.

吴盈鈿·女士
Ms. Lucy W. Wang
(Wu Ying-dian, 1936—)

榮宗敬先生的外孫媳婦、上海老一代紡織專家吳昆生先生的女兒、王雲程先生的兒媳。她1956年赴美國留學，在Mary Washington College大學獲得社會學學士學位，畢業後曾在美國社會福利部門任職。她是位典型的賢妻良母，一生陪同丈夫王建民先生輾轉世界各地，卓有成效地創業，並把兒女都培養成藝術家和實業家。她在海外生活半個多世紀，始終愛穿中國旗袍，尤其在重要的社交場合，總是一身漂亮的旗袍。她的這件淺綠色真絲面料的旗袍套裝，裡面是真絲旗袍，外套採用半透明的蕾絲面料和珠片裝飾，與旗袍色彩相輔相成，雍容華貴，氣場不凡，是出席重要社交活動時的禮服，由她本人親自提供。

Ms. Lucy W. Wang (1936–) is the granddaughter-in-law of Mr. Yung Tsung-chin, the daughter of Mr. Wu Kunsheng, a textile expert in old Shanghai, and the daughter-in-law of Mr. Y. C. Wang. She went to the United States to study in 1956 and received a bachelor's degree in sociology from Mary Washington College. After graduation, she worked in the U.S. social welfare department. She is a traditional wife and mother. She accompanied her husband, Mr. Wang Jianmin, to go around the world throughout her life, starting a successful business, and educating her children to become artists and industrialists. She has lived overseas for more than half a century and has always loved wearing Chinese qipaos. Especially on important social occasions, she always wears a beautiful qipao. Her qipao suit shown in this book is made of light green silk fabric. It has a matching silk qipao inside. The coat is decorated with translucent lace fabric and beads, which complements the color of the qipao. The suit is elegant and luxurious, and has an extraordinary aura. Provided by herself, this dress is certainly fit for attending important social events.

王建民、吳盈鈿夫婦

張樂怡·女士 Ms. Laura C. Soong

(Zhang Le-yi, 1909—1988)

民國時期著名金融家、外交家、民國高級官員宋子文的夫人，江西九江著名商人張謀之的女兒，早年畢業于南京金陵女子大學，後來長期寓居美國。她一生陪同丈夫參加過無數次社會活動，在各種場合她始終身穿中國旗袍。本書刊出她的5件旗袍，都是高檔的旗袍禮服。一件是蕾絲黃白色大菊花面料；第二件是肉色蕾絲面料的珠片繡旗袍夜禮服（胸部以上半透明）；第三件是黃色緞面的暗花珠片繡旗袍；第四件是米色緞面的雙開襟珠繡旗袍；第五件是橙色蕾絲的毛線繡旗袍。這些旗袍均由著名好萊塢華裔演員盧燕的女兒黃漢琪女士（曾是宋子文、張樂怡夫婦的外孫媳婦）保存，經徐芝韻女士聯絡，徐景燦、宋路霞、宋路平在黃漢琪女士家裡拍攝了這組旗袍照片。

Ms. Laura C. Soong (1909–1988) was the wife of T. V. Soong, a famous financier, diplomat, and senior official of the Republic of China in the early 20th century. She was the daughter of Zhang Mouzhi, a famous businessman in Jiujiang, Jiangxi Province. Ms. Laura C. Soong graduated from Ginling College in Nanjing and later lived in the United States. She accompanied her husband to countless important social activities throughout her life, and she always wore a Chinese qipao on various occasions. This book displays five of her qipaos, all of which are high-end elegant dresses. One qipao is made of yellow and white large chrysanthemum lace fabric; the second is a sequin-embroidered qipao evening dress made of flesh-colored lace fabric (translucent above the chest); the third is a yellow satin one with dark floral sequin embroidery; the fourth is a beige satin double-fronted and bead-embroidered qipao; the fifth is an orange lace woolen-embroidered one. These qipaos were preserved by Ms. Lucia Hwang Gordon who is the daughter of the famous Hollywood Chinese actress Lisa Lu. Ms. Lucia Hwang Gordon was formerly the granddaughter-in-law of Ms. Laura C. Soong. After contacted by Ms. Carolyn Hsu Balcer, Ms. Jeanette Zee, Ms. Song Luxia, and Mr. Song Luping took these qipao photos at Ms. Lucia Hwang Gordon's home.

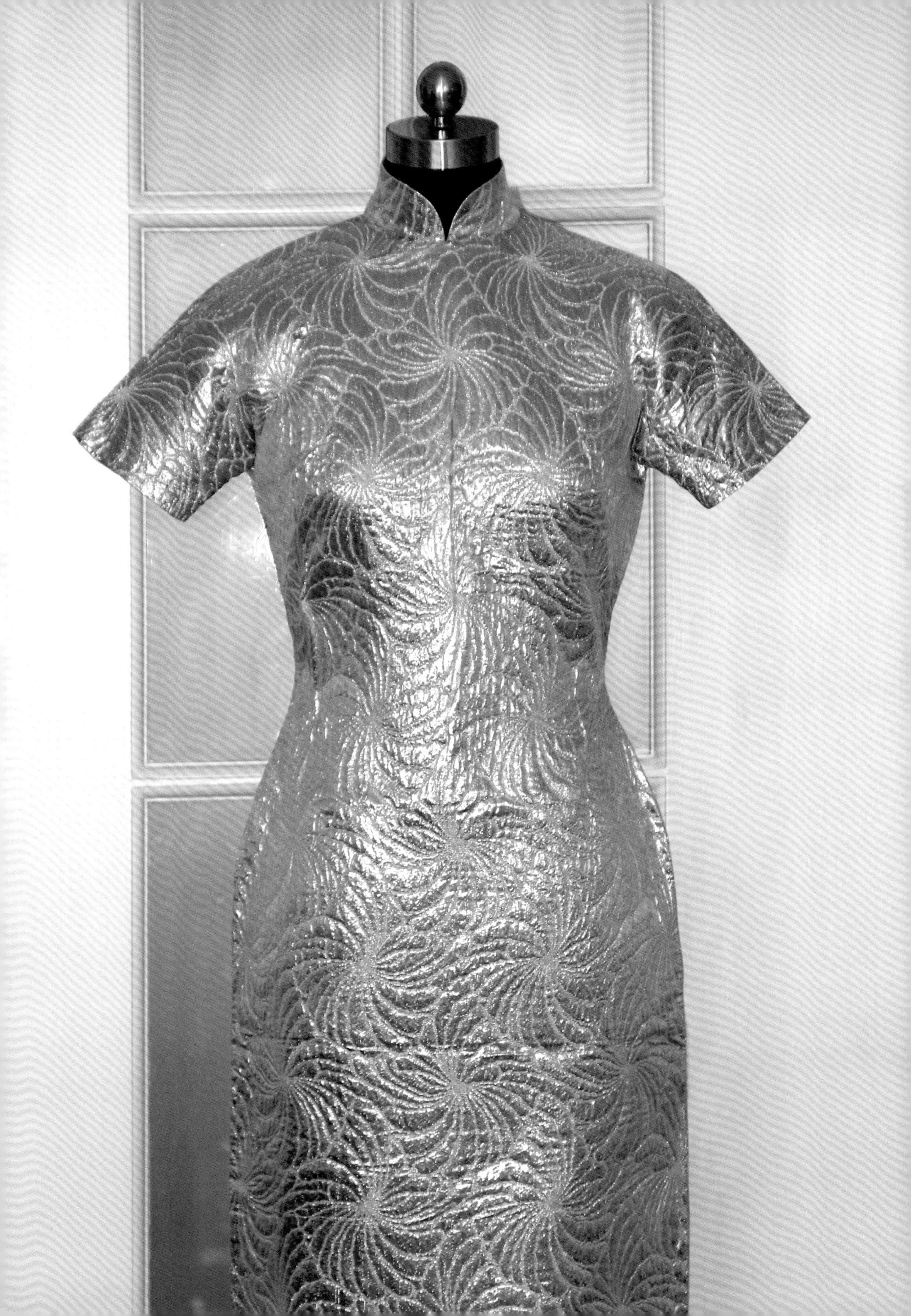

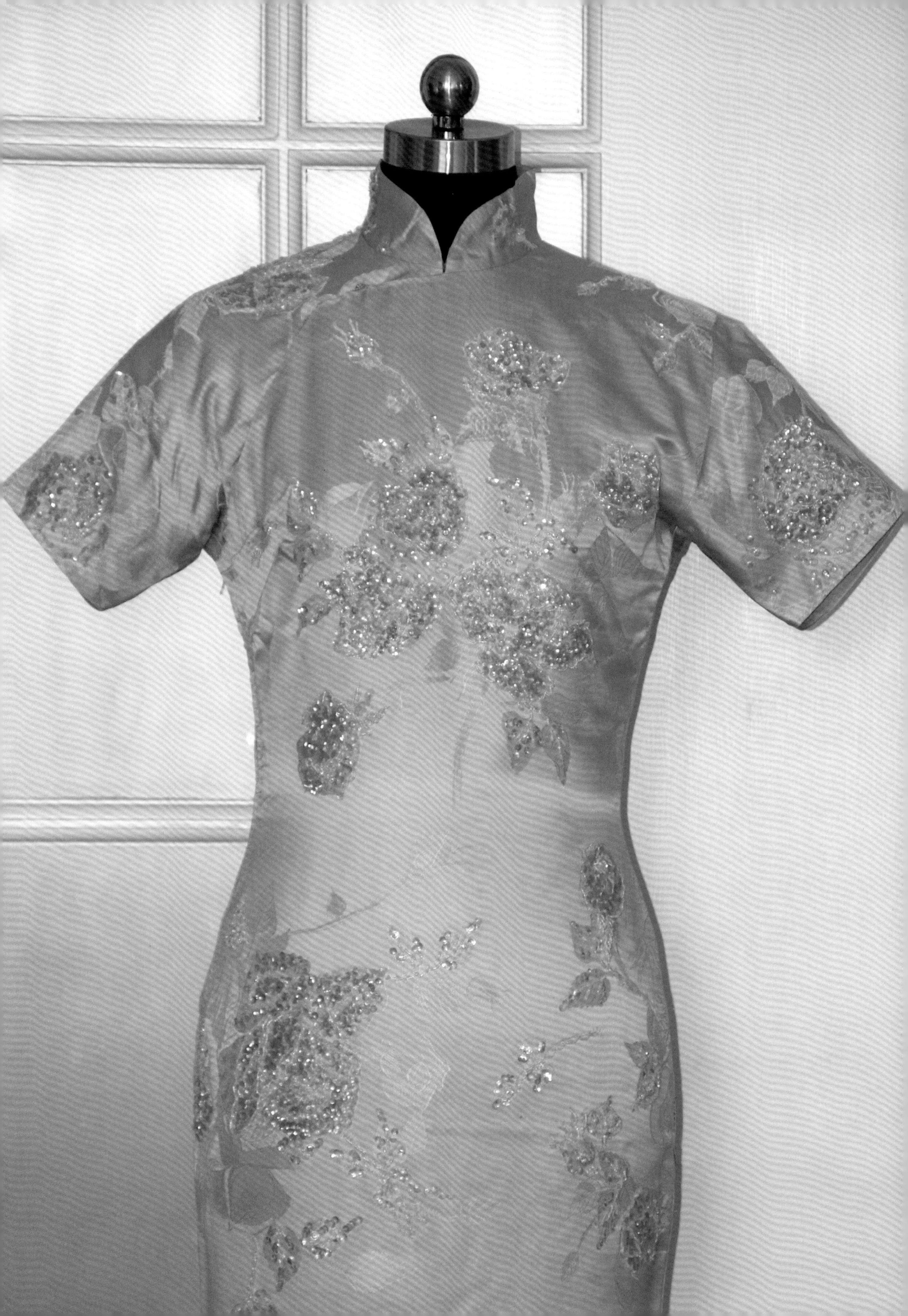

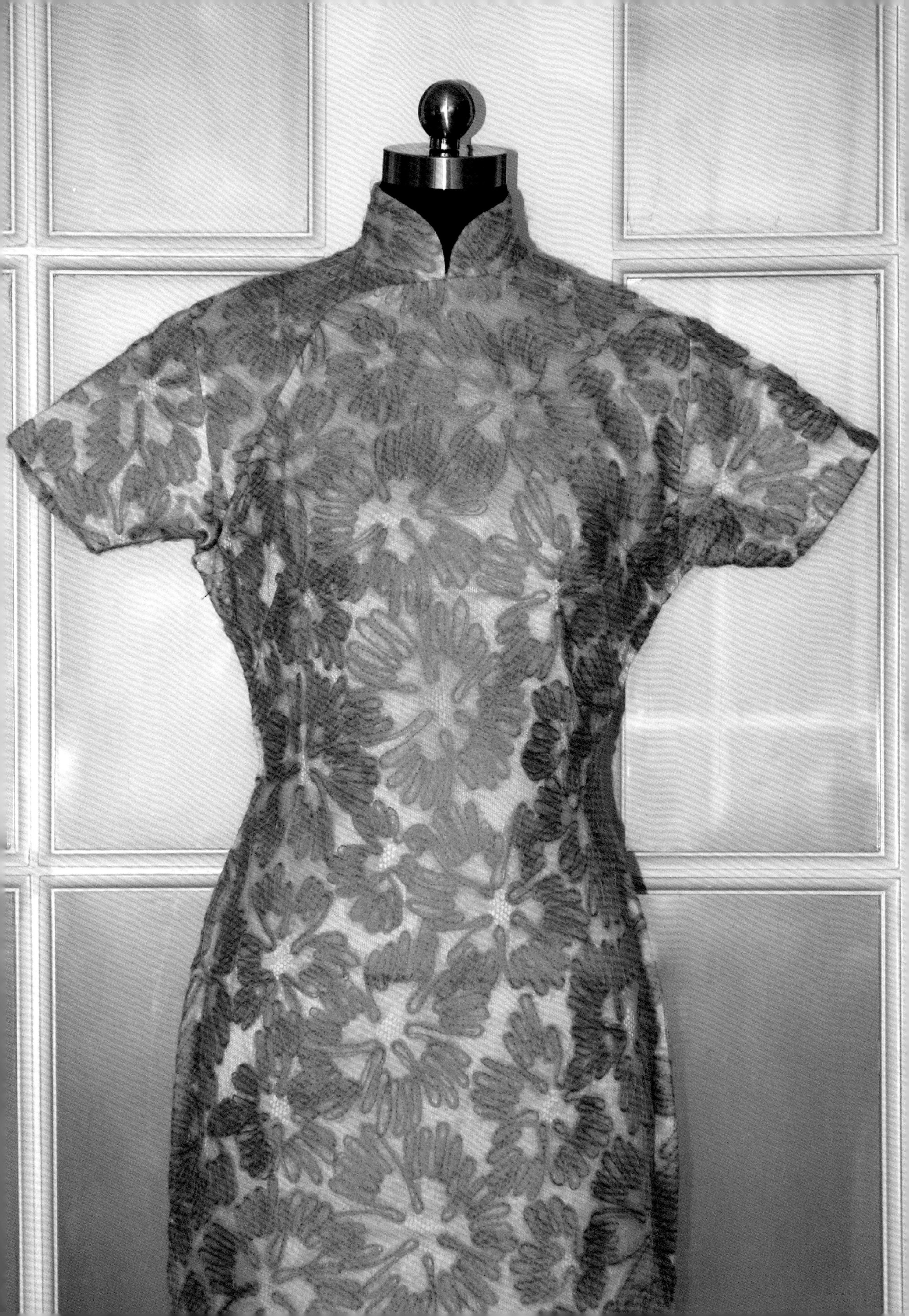

張鐘秀・女士

Ms. Zhang Zhongxiu

(Zhang Zhong-xiu, 1887–1969)

蘇州人，蘇州大鹽商張履謙的長孫女。她16歲嫁到上海靜安寺路上的盛公館，成爲晚清洋務巨擘盛宣懷的弟媳婦，即盛善懷的夫人。老人家知書達理，性格爽直，遇事果斷，很有丈夫氣。丈夫去世後，面對家道中落、世事紛亂的民國萬象，她從不畏懼，坦然地獨自持家，把女兒盛范頤培養成人，並在抗戰期間與女兒一起，跟女婿孫蔚青前往福建參加抗日工作。這件黑色旗袍是她中年時代穿的，款式傳統，做工細膩，長期由她的外孫孫世仁、外孫女孫世瑾保管，2022年捐贈給上海老旗袍珍品館。

Ms. Zhang Zhongxiu (1887–1969) was a native of Suzhou and the eldest granddaughter of Zhang Lüqian, a great salt merchant in Suzhou. At the age of 16, she married into the Sheng's Mansion on Jing'an Temple Road in Shanghai and became the sister-in-law of Sheng Hsuan-huai, a tycoon of Westernization in the late Qing Dynasty, that is, the wife of Sheng Shanhuai. The old lady was well-educated, straightforward in making decisions in tough situations, and therefore very strong-minded. After her husband's death, Ms. Zhang was never afraid of the decline of her family and the chaos of the Republic of China. She took care of the household calmly and raised her daughter Sheng Fanyi to adulthood. During the War of Resistance Against Japanese Aggression, she went to Fujian with her daughter and son-in-law Sun Weiqing to participate in the work against Japan. This black qipao was worn by her in her middle age, with a traditional style and exquisite workmanship. It had been kept by her grandson Sun Shiren and granddaughter Sun Shijin for a long time, till it was donated to Shanghai Old Qipaos Collection Museum in 2022.

趙一荻・女士

Ms. Edith Chao

(Zhao Yi-di, 1912—2000)

晚年的趙四小姐和張學良將軍

趙四小姐是張學良將軍的紅粉知己、現代中國最具傳奇的堅強女性之一。她與張學良將軍牽手36年後才有了正式夫人的名份，1964年與張學良將軍在臺灣結婚。她把一生都奉獻給了張將軍，陪伴張將軍達72年，與其共同經歷了現代中國雲譎波詭的政治風雲，包括張將軍被監禁的漫長歲月，被譽爲愛的化身。這件紅色絲麻面料的旗袍套裝款式簡潔，幾何暗紋提花，呈現多層次的花卉紋樣，非常鮮豔，是她在張學良將軍百歲壽慶活動中穿的禮服，最近由她哥哥趙燕生先生的孫女趙荔女士從夏威夷帶回上海，捐贈給上海老旗袍珍品館。

Ms. Edith Chao (Zhao Yi-di) (1912–2000) was a lady of a distinguished and noble family who was considered one of the most legendary and courageous women of modern China. She was the greatest love of Young Marshal Chiang Hsueh-liang and dedicated her whole life to him. After living together with him for 36 years, she finally became his official wife in 1964 in Taiwan. They were together for 72 years, including the long period of Young Marshal's imprisonment. Therefore, Edith was hailed as the "Embodiment of Love". This qipao from her vast wardrobe is made of red silk and linen fabric, simple in style and bright in color, with dark geometric jacquard and multi-level floral patterns. It was the dress she wore at Young Marshal Chiang Hsueh-liang's centenary celebration, and was recently brought to China by her brother Lonny Chao's granddaughter Zhao Li from Hawaii and donated to Shanghai Old Qipaos Collection Museum.

周煉霞·女士
Ms. Zhou Lianxia
(Zhou Lian-xia, 1906—2000)

江西吉安人，字紫宜，號螺川，室名螺川詩屋，後以筆名煉霞行世，上海著名女畫家。她14歲始先後師從尹和白、鄭凝德學畫，17歲起師從朱古微學詩，後又從蔣梅笙學詞。周煉霞品貌雙全，氣質高雅，擅長繪畫、書法、詩詞，時有『金閨國士』之譽。她繪畫精于仕女和花卉，畫風有唐人韻致；詩詞多佳句，著有《螺川韻語》《學詩淺說》等。民國時期曾發起成立中國女子書畫會，建國初上海籌建中國畫院，其爲第一批進入畫院的9位女畫家之一。這件黑色絲絨旗袍是周煉霞晚年訂製的，後由青年收藏家吳崢嶸先生收藏，繼而捐贈給上海老旗袍珍品館。

Zhou Lianxia (1906–2000) was born in Ji'an, Jiangxi Province. Her courtesy name was Ziyi and nickname was Luochuan. Her study was called "Luochuan's Poetry Room". She later went on to live under her pen name Lianxia. Ms. Zhou was a famous female painter in Shanghai. She began studying painting with Yin Hebai and Zheng Ningde at the age of 14, poetry with Zhu Guwei at the age of 17, and later poetry with Jiang Meisheng. Ms. Zhou Lianxia was beautiful and elegant, good at painting, Chinese calligraphy and poems. For that she was quite often called the "Golden Lady". Her painting style had the charm of Tang Dynasty since she was especially good at painting graceful ladies and beautiful flowers. While she composed many beautiful poems, she was also the author of two books: *Luochuan's Rhymes* and *A Brief Introduction to Poetry*. During the Republic of China, she initiated to found the Chinese Women's Calligraphy and Painting Association. As Shanghai prepared to establish a Chinese Painting Academy during the early days of New China, she was one of the first nine female painters to be admitted. This black velvet qipao was customer-made as ordered by Ms. Zhou Lianxia in her later years. It was later owned by young collector Mr. Wu Zhengrong and then donated by him to Shanghai Old Qipaos Collection Museum.

周賽男 · 女士
Ms. Zhou Sainan
(Zhou Sai-nan, 1923—)

上海人，工程師，1945年畢業于聖約翰大學土木工程系，是該系唯一的女生。大學畢業後，她先後在上海華東設計院、香港司徒惠事務所任職，曾參與1970年日本世博會香港館的設計，20世紀80年代，又參加了廣州花園酒店等大型公共建築的設計。她畢生從事建築設計工作，倍受業界人士尊敬。她的這件紫色菱形碎花的旗袍套裝，面料獨特，是在紫色幾何紋樣的底布上增加一層亮片，典雅大氣，自帶光芒。外套爲翻領闊袖，是典型的職業女性喜愛的旗袍套裝款式，由她的女兒鄭璐璐女士捐贈給上海老旗袍珍品館。

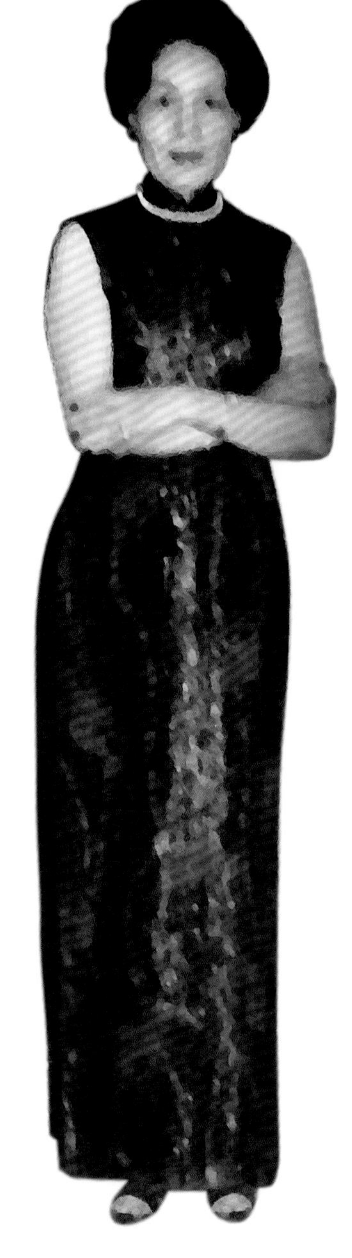

Ms. Zhou Sainan (1923-) is a native of Shanghai and an engineer. She graduated from the Civil Engineering Department of St. John's University Shanghai in 1945 and is the only female student in the department. After graduating from university, she worked at Shanghai East China Design Institute and Hong Kong Situ Hui Office respectively. As a fine civil engineer, she participated in the design of Hong Kong Pavilion at the World Expo 1970 in Japan. During the 1980s, she participated in the design of large public buildings such as Guangzhou Garden Hotel. She has been engaged in architectural design work all her life and is highly respected by people working in the civil engineering field. Her purple diamond-shaped floral qipao suit in this book has a unique fabric, with a layer of sequins to the purple geometric pattern base, which is elegant and bright. The coat has wide lapels and sleeves, which is of a typical qipao suit style favored by professional women. This qipao was donated to Shanghai Old Qipaos Collection Museum by her daughter, Ms. Zheng Lulu.

周靜慧・女士
Ms. Mabel C. Chang
(Zhou Jing-hui, 1920—)

周靜慧女士（中）與貝聿銘夫婦

民國元老、民國時期曾三任杭州市市長的周象賢的女兒，年輕時赴美留學，1948年考入聯合國工作，任聯合國會議事務部會議規劃服務科組長，服務期長達32年，現居美國紐約。她在美國生活近80年，始終喜歡穿中國旗袍。本書刊出她的兩件旗袍，一件是真絲印花嵌銀絲的無袖旗袍，另一件是米色軟緞的珠繡旗袍套裝，均製作精細，圖案靚麗，令人眼前一亮。

Ms. Mabel C. Chang (1920-) is the daughter of Z. Y. Chow, a veteran of the government who served as the mayor of Hangzhou three times during the Republic of China. She went to the United States to study when she was young. In 1948, she was admitted to the United Nations and served as the Director of the Conference Planning Services Section under the United Nations Conference Affairs Department for 32 years. She currently lives in New York. Although Ms. Mabel C. Chang has been living in the United States for nearly 80 years, she always like to wear traditional Chinese qipaos. This book displays two of her qipaos: one is a silk-printed sleeveless qipao, woven with silver threads, and the other is a beige soft satin bead-embroidered qipao suit. Both are exquisitely made and have beautiful patterns that are really eye-catching.

鐘曙華 · 女士
Ms. Zhong Shuhua
（Zhong Shu-hua, 1911—2010）

原名鐘琳英，出生於上海浦東召稼樓附近的『鐘家堂』，一個教育世家。她3歲喪父，在家族的祠堂裡接受私塾教育，19歲嫁給進步青年陸詒。1932年淞滬抗戰爆發時，陸詒從事戰地新聞採訪，後來成爲著名的戰地記者。鐘曙華女士跟隨夫君走南闖北，戰爭年代顛沛流離，曾流亡香港、新加坡、緬甸等地。這件藍色真絲提花旗袍，據說就是在香港時訂製的，是隨夫君陸詒出席社交場合時穿的服裝，樣式簡潔，樸素無華，由她的兒子陸良年、孫女陸齊虹捐贈給上海老旗袍珍品館。

陸詒、鐘曙華夫婦

Ms. Zhong Shuhua (1911–2010), formerly known as Zhong Linying, was born into a family of scholars near Zhaojialou, Pudong, Shanghai. Her father died when she was three years old. She received private education in the family's ancestral hall. At age 19, she married a progressive-minded young man whose name was Lu Yi. When the Battle of Songhu (Shanghai) broke out in 1932, Mr. Lu Yi was engaged in battlefield news reporting and later became a famous war correspondent. Ms. Zhong then followed her husband all over the country. During those years, she can be said as homeless since she had lived in Hong Kong, Singapore, Myanmar and many other places. This blue silk jacquard qipao is said to have been custom-made in Hong Kong. It was worn by Ms. Zhong when attending social occasions with her husband. The style is simple and unpretentious. This qipao was donated to Shanghai Old Qipaos Collection Museum by her son Lu Liangnian and granddaughter Lu Qihong.

附录

芝莲福文化发展有限公司复制名媛旗袍作品

後記

在這本書中，細心的讀者會發現，有相當一部分旗袍是珠繡旗袍，比如：邊陳之娟博士、陳香梅女士、鄧戴月華女士、賀寶善女士、林鄭月娥女士、宋胡靜君女士、吳盈鈿女士、張樂怡女士、周賽男女士、周靜慧女士，她們的旗袍都以漂亮的珠繡旗袍著稱。附錄中芝蓮福文化發展有限公司複製的名媛旗袍，也都是珠繡旗袍。

探究這些典雅華貴的珠繡旗袍的背後故事，我們漸漸明白了，珠繡旗袍不僅面料考究、五光十色、做工複雜、價格昂貴，而且，是名媛們在重要場合的穿著禮服，比如結婚典禮、慶生聚會、外事活動，或重要的社會活動，珠繡旗袍常常是名媛服裝的首選。

由此可知，海派旗袍的巔峰之作堪推珠繡旗袍。“珠光寶氣”的服飾效果，除了佩戴珠寶首飾之外，還可以以這種形式得以“閃亮登場”。儘管珠繡旗袍沒有在中國大陸風行過，但是在香港曾經大紅大紫，高峰期是20世紀50年代到90年代。世界各地的中國閨秀，都喜歡去香港訂做旗袍，尤其是珠繡旗袍，一定是在香港訂做，因為那時的香港，彙集了一批從上海遷過去的老裁縫，出現了一批卓有特色的高端旗袍店。

我們獲贈的張樂怡女士、嚴幼韻女士、蔣士芸女士、張幼儀女士、華若芸女士、姚翠棣女士、吳盈鈿女士、邊陳之娟博士、賀寶善女士、姜何琴霞女士、沙聶惠珍女士、邵婉琴女士、周賽男女士等人的珠繡旗袍，都是在香港訂做的，可知那時的香港，已然是海派旗袍的大本營，“上海裁縫”在香港旗袍界，贏得了“天花板”級別的口碑。

前不久，我們有幸採訪到的“張水法時裝設計公司”就是一例。我們在榮家兩位老太太的多件旗袍上，發現了該公司的商標。現在知道，該公司在香港創辦於1946年，創辦人張水法先生是上海浦東人，抗戰勝利後就赴香港創業，現已去世，目前的掌門人是他的兒子兒媳張德忠、忻靄苓夫婦，以及孫子張正年先生。住在香港的老一代上海名媛，對他們幾乎無人不知……從某種意義上說，是張水法這樣一批“上海裁縫”，把海派旗袍從上海成功地“登陸”到香港。這個故事似乎在召喚我們，應該再去香港，進一步發掘上海與香港互動的“雙城記”，尤其是旗袍故事，因為這是目前，旗袍文化所“失憶”的一部分啊！

書中絕大部分旗袍，現存上海巨鹿路上的“元芳閣”。這是由香港范元芳老師出資贊助，我們租的上海老旗袍珍品館的館舍。目前，“元芳閣”裡不僅存放了數百件名媛旗袍，還是旗袍姐妹組織小型研討和交流活動的場所，受到不少旗袍團隊和沙龍組織的歡迎。在此，我們向范元芳老師致以衷心的感謝！

這些美麗的旗袍如今能夠薈萃於一編，見證旗袍文化的發展，是眾多長輩和朋友們鼎力相助的結果。前香港特首林鄭月娥女士在百忙之中，仍關注旗袍文化的發展。她的旗袍來到上海，使眾多的旗袍姐妹歡欣鼓舞，爭相觀賞。她的朋友邱海斌先生熱心促成了這件好事。

原上海百樂門創辦人顧聯承先生的孫子顧家璉先生，在經營企業的同時，還擔任香港湖州南潯同鄉會的會長。他大力助推滬港兩地旗袍文化的交流，不僅將其母親陳天真老人留下的 15 件旗袍，捐贈給我們上海老旗袍珍品館，還組織我們參與香港旗袍組織的聯誼活動，並數次親自駕車上山，帶我們前往香港名媛的家裡，觀賞和借用旗袍。我們不能不為之深深感動。

年已 93 歲高齡的吳順昇老先生是上海“老克勒”，他與齊如山老人的外甥女賀寶善女士是老朋友。賀寶善女士去世後，她的旗袍由其女兒姚詠蓓女士收藏。姚詠蓓女士遠在英國，吳老先生就不遠萬里“遙控指揮”，終於使賀寶善女士的 3 件美麗旗袍，落戶上海。

我們還要深深感謝榮周淑霞女士，孫國平先生，邊陳之娟博士，鄧戴月華女士，簡宋麗迪女士，劉蓮華女士，黃漢琪女士，吳盈鈿女士，周靜慧女士，盛承慧女士，趙荔女士，吳崢嶸先生，朱曾滸、高麗霞夫婦，葉世棟、李麗夫婦，王麗瑞女士，孫世仁、孫世瑾兄妹，鄭璐璐女士，陸良年、陸齊虹父女，陳慧萍女士，樓國平女士。

我們還要感謝本書編委會的全體成員及出版社的編輯們，在他們的慷慨贊助及辛勤努力下，才有這本旗袍書的完美呈現。這一切，都將鞭策我們，把傳承旗袍文化的事業，做得更好。

謝謝大家。

宋路霞／周鐵芝

2024 年 4 月

Acknowledgment

Careful readers may have noticed that a considerable portion of the qipaos in this book is embroidered with beads. For example, the qipaos worn by Dr. Delia Chen Chi Kuen Pei, Mrs. Anna Chan Chennault, Ms. Dora Tang Tai Yuet Wah, Ms. He Baoshan, Ms. Carrie Lam Cheng Yuet-ngor, Ms Soong Hu Jingjun, Ms. Lucy W. Wang, Ms. Laura C. Soong, Ms. Zhou Sainan and Ms. Mabel C. Chang are all famous for such a feature. The qipaos replicated by Shanghai Zhi Lian Fu Culture Co., Ltd. as shown in the appendix are also bead-embroidered ones.

Exploring the stories behind these elegant and luxurious qipaos, we have gradually known that bead-embroidered qipaos not only feature exquisite fabrics, colorful designs, complex craftsmanship and high prices, but also are the attire worn by older generation ladies on important occasions such as wedding ceremonies, birthday parties, foreign affairs events, or important social events. For the ladies, they are often the preferred clothing.

Such being the case, it can be said that the pinnacle of Shanghai-style qipaos is a bead-embroidered one. In addition to wearing jewelry, the "jewel-like" clothing effect can also be "brilliantly shown" in this form. Never popular in Chinese Mainland, bead-embroidered qipaos were once very popular in Hong Kong, with the peak period from the 1950s to the 1990s. Chinese ladies from all over the world would like to go to Hong Kong to order qipaos, especially bead-embroidered ones. These clothes had to be custom-made in Hong Kong, because at that time, a number of Shanghai tailors gathered in Hong Kong, and a number of distinctive high-end qipao stores also emerged there.

The bead-embroidered qipaos we have received, including those worn by Ms. Laura C. Soong, Ms. Juliana Y. Koo, Ms. Aileen C. Pei, Ms. Chang Yu-i, Ms. Hua Ruoyun, Ms. Sally Y. Wang, Ms. Lucy W. Wang, Dr. Delia Chen Chi Kuen Pei, Ms. He Baoshan, Ms. Helen H. Chiang, Ms. Jean N. Sa, Ms. Jean S. Hsu and Ms. Zhou Sainan, were all custom-made in Hong Kong. It can be seen that Hong Kong was already the headquarters of Shanghai-style qipaos at that time, and Shanghai tailors won a "ceiling" level reputation in the qipao industry in Hong Kong.

One example is that not long ago, we had the privilege of interviewing Zeepha Chiang Fashion

Design Company. We discovered the company's trademark on many of the qipaos worn by two elderly ladies in the Yung family. It is now known that the company was founded in Hong Kong in 1946, and its founder, Mr. Zeepha Chiang, was from Pudong, Shanghai. After the victory of The War of Resistance Against Japanese Aggression, he went to Hong Kong to start a business and has now passed away. The current bosses are his son Zhang Dezhong, daughter-in-law Xin Ailing, and grandson Zhang Zhengnian. Almost each of the older generation Shanghai ladies living in Hong Kong knows about them ... In a sense, it is a group of "Shanghai tailors" represented by Zeepha Chiang who has successfully brought Shanghai-style qipaos from Shanghai to Hong Kong. This story seems to be calling us to go to Hong Kong again and further explore the interaction between Shanghai and Hong Kong, especially the qipao story, because this is a part of qipao culture that has lost its memory.

The vast majority of the qipaos in the book is stored at Yuen Fong's Pavilion on Julu Road in Shanghai, the seat of Shanghai Old Qipaos Collection Museum sponsored by Ms. Fan Yuen Fong from Hong Kong and rented by us. Yuen Fong's Pavilion not only stores hundreds of famous qipaos, but also is a place for qipao sisters to organize small seminars and cultural exchanges, which has been welcomed by many qipao groups and salons. We would like to take this opportunity to extend our heartfelt thanks to Mr. Fan Yuen Fong.

The fact that these beautiful qipaos can now be gathered in this book and witness the development of qipao culture is the result of the great assistance of many elders and friends. The former Chief Executive of Hong Kong, Ms. Carrie Lam Cheng Yuet-ngor, is concerned about the development of qipao culture in spite of her busy work. The arrival of her qipaos in Shanghai has made many qipao sisters happy and eager to appreciate. Her friend, Mr. Qiu Haibin, has enthusiastically contributed to this good thing.

Mr. Gu Jialian, the grandson of Mr. Gu Liancheng (the former founder of Shanghai Paramount), not only runs a business but also serves as the president of Association of Fellow Townsmen from Nanxun District of Huzhou City in Hong Kong. Vigorously promoting the exchanges of qipao culture between Shanghai and Hong Kong, he has not only donated the 15 qipaos left by his mother, Ms. Chen Zhenzhen, to Shanghai Old Qipaos Collection Museum, but also organized us to participate in the activities organized by Hong Kong qipao organizations. He personally drove us up Victoria Peak several times to visit, appreciate and borrow qipaos from older generation Shanghai ladies living in Hong Kong. We cannot help but be deeply moved by it.

Mr. Otto Wu, who is 93 years old now, is known as a "Shanghai-style old gentleman". He and Ms. He Baoshan, the niece of Mr. Qi Rushan, are old friends. After Ms. He passed away, her qipaos were collected by her daughter, Ms. Yao Yongbei. Since Ms. Yao was in the UK, through remote command, Mr. Wu finally helped Ms. He's three beautiful qipaos settle in Shanghai.

We would also like to express our sincere gratitude to Ms. Yung Zhou Shuxia, Mr. Sun Guoping, Dr. Delia Chen Chi Kuen Pei, Ms. Dora Tang Tai Yuet Wah, Ms. Lydia Soong Kan, Ms. Rio L. Chiang, Ms. Lucia Hwang Gordon, Ms. Lucy W. Wang, Ms. Zhou Jinghui, Ms. Sheng Chenghui, Ms. Zhao Li, Mr. Wu Zhengrong, the couple of Mr. Zhu Zenghu and Ms. Gao Lixia, the couple of Mr. Ye Shidong and Ms. Li Li, Miss Lily Falzon Wong, brother and sister Mr. Sun Shiren and Ms. Sun Shijin, Ms. Zheng Lulu, father and daughter Mr. Lu Liangnian and Ms. Lu Qihong, Ms. Chen Huiping, and Ms. Lou Guoping.

We would also like to thank all the members of the editorial board and the editors of the publishing house for their generous sponsorship and hard work, which led to the perfect presentation of this book. All of this will inspire us to do a better job in inheriting qipao culture.

Thank you all.

Song Lu-xia, Zhou Tie-zhi

April, 2024

圖書在版編目（CIP）數據

滬港名媛旗袍寶鑒 / 宋路霞，徐景燦，周鐵芝編著．— 上海：上海科學技術文獻出版社，2024

ISBN 978-7-5439-9170-5

I. TS941.717.8

中國國家版本館 CIP 數據核字第 20241XF256號

責任編輯　于學松
特約編輯　陳寧寧
裝幀設計　儲　平

滬港名媛旗袍寶鑒
宋路霞 徐景燦 周鐵芝 編著　王逸 俞晨瑋 譯　宋路平 攝影

出版發行　上海科學技術文獻出版社
地　　址　上海市淮海中路1329號4樓
郵政編碼　200031
經　　銷　全國新華書店
印　　刷　上海中華商務聯合印刷有限公司
開　　本　889×1194　1/16
印　　張　9.5
版　　次　2024 年 8 月第 1 版　2024 年 8 月第 1 次印刷
書　　號　ISBN 978-7-5439-9170-5
定　　價　198.00 圓
http://www.sstlp.com